W9-AWQ-240

What Do You Like?

Copyright © Gareth Stevens, Inc. All rights reserved.

Developed for Harcourt, Inc., by Gareth Stevens, Inc.
This edition published by Harcourt, Inc., by agreement with Gareth Stevens, Inc. No part of this publication may be reproduced or transmitted in any form or by any means, electronic or mechanical, including photocopy, recording, or any information storage and retrieval system, without permission in writing from the copyright holder.

Requests for permission to make copies of any part of the work should be addressed to Permissions Department, Gareth Stevens, Inc., 330 West Olive Street, Suite 100, Milwaukee, Wisconsin 53212. Fax: 414-332-3567.

HARCOURT and the Harcourt Logo are trademarks of Harcourt, Inc., registered in the United States of America and/or other jurisdictions.

Printed in the United States of America

ISBN 13: 978-0-15-360178-1
ISBN 10: 0-15-360178-7

1 2 3 4 5 6 7 8 9 10 039 16 15 14 13 12 11 10 09 08 07

What Do You Like?

by Jennifer Marrewa

Photographs by Patrick Espinosa

Harcourt

SCHOOL PUBLISHERS

Chapter 1:
Where Should We Have the Party?

It is June, and school is almost over. The children in the neighborhood are planning a summer block party. Everyone who lives in the apartment buildings will be invited.

The children will meet at Emma's apartment after school. Her grandmother will help the children plan the party.

The neighborhood children plan the summer block party.

First, the children will choose where to have the party. They have three ideas. They could have the party at one of their apartments. The courtyard between the apartment buildings is another idea. They could also have the party at the neighborhood park.

Clancy lists the three choices on a sheet of paper. He says, "We need to pick one place to have the block party."

The children write their names on the sheet of paper. They vote for the best place to have the block party.

Where Should We Have the Block Party?

Apartment	Courtyard	Park
	Ruben	Sam
	Gloria	
	Mai	

Everyone voted on where to hold the block party. Most of the children think the courtyard is the best place to have the party. There will be lots of room to play in the area between the apartment buildings. The party will be close to all the neighbors, too.

Where Should We Have the Block Party?

Apartment	Courtyard	Park
Clancy	Ruben	Sam
	Gloria	Kim
	Mai	
	Tim	
	Sophie	
	Cami	

Chapter 2:
What Should We Do at the Party?

    Next, the children will decide what kind
of food to have at the party.

    Should they have a big bowl of popcorn?
What about pretzels or carrots with
vegetable dip? Sophie thinks string cheese
and sliced apples are tasty foods to have at
the block party.

Jesse writes the ideas on his paper. Everyone votes for a favorite food, and Jesse makes tally marks on his tally table to record the votes.

What would the neighbors like to eat? The children will ask some neighbors, and then they will add this data to the tally table.

Food for the Summer Block Party	
Popcorn	II
Pretzels	I
Carrots	I
Cheese	III
Apples	II

Lemonade, water, milk, and juice could be served at the party.

Emma's grandmother says that they should have drinks at the block party. She has a pitcher and some plastic cups the children can use that day.

Ruben thinks lemonade or water would be good to serve on a warm day. Some people might want to drink milk or orange juice.

Gloria lists the drink choices on a sheet of paper. She asks everyone to vote for a favorite drink. Gloria draws a dot for each kind of drink.

Lemonade gets the most votes. Gloria will talk to her neighbors to find out whether they like lemonade, too.

Drinks for the Summer Block Party

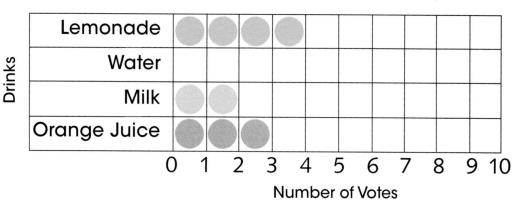

Musical chairs would be a fun game to play at the party.

There should be a game at the party that everyone can play. The children name some of their favorite games. Musical chairs and freeze tag are fun games to play outside. Everyone could join in a treasure hunt for a prize hidden in the courtyard.

Ruben makes a list of game ideas on a sheet of paper. The children vote for their favorite games. He records this data on his grid paper, and then he makes a bar graph showing the data. The children vote to play musical chairs at the block party.

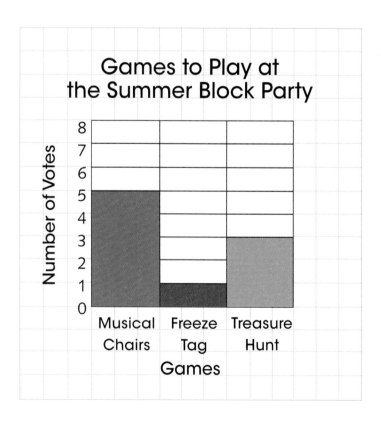

Games to Play at the Summer Block Party

Chapter 3:
Which One Is Your Favorite?

The children have chosen their favorite food, drink, and game. Will their neighbors like the same things? The children will divide into groups and talk to some of their neighbors. The children will ask them to vote for a favorite food, drink, and game, too.

The children ask the neighbors to vote for their favorite food, drink, and game.

After talking to some neighbors, the children will meet at Emma's apartment again. They will add the data from the neighbors to the graphs. Then, they will finish planning the block party.

Mrs. Hernandez votes for musical chairs.

Mrs. Hernandez is Jesse's mom. She thinks musical chairs would be the most fun for everyone to play. Ruben records her vote on his graph. Now there are six votes for musical chairs.

Mr. Orta votes for sliced apples.

Mr. Orta lives next door to Sophie. He thinks sliced apples would be a good snack for the party. Jesse makes a tally mark next to *sliced apples* on his tally table. Mrs. Orta votes for sliced apples, too. Now there are four votes for sliced apples.

Chapter 4:
The Results Are In!

The children spend the afternoon talking to their neighbors in the apartments. The neighbors are all happy to hear about the plans for a party. After gathering data, the children return to Emma's apartment. They look at the data to see the favorite food, drink, and game.

Gloria shows the graph she made of drink choices. Only one person thinks it would be a good idea to serve water. Lemonade has the most votes. Eight people said lemonade is their favorite drink. The children will have lemonade at the block party.

Drinks for the Summer Block Party

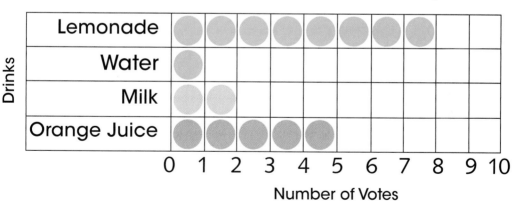

Jesse shows the tally table he made for food. Two foods have the same number of votes. String cheese and sliced apples each have four votes. They will serve both foods at the block party.

Food for the
Summer Block Party

Popcorn	III
Pretzels	I
Carrots	II
Cheese	IIII
Apples	IIII

Ruben shares the bar graph he made for games. The longest bar shows the children which game got the most votes. The longest bar on the graph is for musical chairs. The children think everyone will like that game.

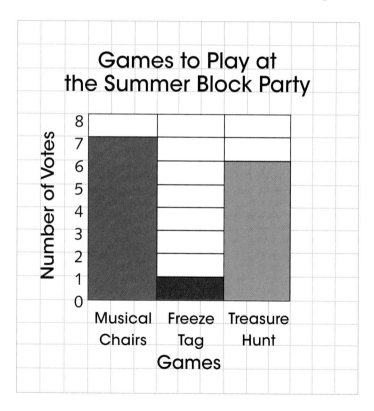

Games to Play at the Summer Block Party

The plans for the summer block party are complete.

The plans for the party are done. The children will serve string cheese and sliced apples. They will also have lemonade to drink at the block party. They will set up a circle of chairs in the courtyard and ask everyone to play musical chairs.

Now Sophie and Ruben get to work making invitations for the party. The children think many of the neighbors will want to go to their summer block party. Everyone will have fun!

Come to the Summer Block Party in the Courtyard Next Saturday!

Glossary

bar graph a drawing that uses bars of different lengths to show different numbers or amounts

data facts or information

result an outcome

tally the mark in a tally table that stands for one thing